griddlers
Logic Puzzles

M000275304

Black and White

Volume 1

Smart Things Begin With Griddlers.net

Griddlers Logic Puzzles: Black and White Vol. 1

Published by: Griddlers.net
a division of A.A.H.R. Offset Maor Ltd

Author: Griddlers Team
Compiler: Rastislav Rehák
Cover design: Elad Maor
Contributors: abrek, airam, alecmyers, alibaba, amam100, Amit11, apmega, arcadedweller, Ardna, aya, BenMiff, coachkelly, DinaGreen, dromidror, eleonor, elimaor, esra, ferrari, fertu, galisand, Glucklich, Gothic56, griddlers_books, harmless_52, Heracleum, hi19hi19, hibrahimozer, HSpring, Irin4ik, jajongma, JESlovesIVO, jfred99, karoline, kissyrich, laurel, ld5, liri748, LuB, luvlaketahoe, MauiTiger, MDE001, Mini-Fayer, mko, mysza9, nasa17, Ne_Plus_Ultra, nico5038, nimrod, NOIRAUD, nutrija, oko, op1, osher32, Oskar, poj, Ra100, Rainbow15, razzmatazz, redhead64, rehacik, Rianne1992, sandyeggan, skybreezes, solo222, starch, stijn98, stumpy, syucesan, tjbpm, twoznia, ulka, vargben, xiayu, xxLadyJxx, ylletrollets, yolinde, Zaba258, zjmonty, Zodrex, zolesz

ISBN: 978-9657679005

More information:
Email – team@griddlers.net
Website – http://www.griddlers.net

Definition

Griddlers, also known as Paint by Numbers or Nonograms, are picture logic puzzles in which cells in a grid have to be colored or left blank according to numbers given at the side of the grid to reveal a hidden picture. In this puzzle type, the numbers measure how many unbroken lines of filled-in squares there are in any given row or column. For example, a clue of "4 8 3" would mean there are sets of four, eight, and three filled squares, in that order, with at least one blank square between successive groups.

These puzzles are often black and white but can also have some colors. If they are colored, the number clues will also be colored in order to indicate the color of the squares. Two differently colored numbers may or may not have a space in between them. For example, a black four followed by a red two could mean four black spaces, some empty spaces, and two red spaces, or it could simply mean four black spaces followed immediately by two red ones.

There are no theoretical limits on the size of a nonogram, and they are also not restricted to square layouts. (From Wikipedia, the free encyclopedia)

Basic Rules

Each clue indicates a group of contiguous squares of like color.

Between each group there is at least one empty square.

The clues are already in the correct sequence.

Groups of different colors or different triangles may or may not have empty squares between them. Triangle without a number counts as 1.

Step by Step Example

Following example will show you how to solve easy puzzle. Mark empty squares by dot. We suggest to use marking especially for difficult puzzles.

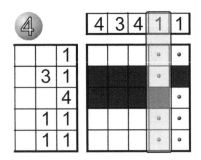

Row 2: Clues (3,1), with 1 empty square (background color) between them, add up to the 5 available squares.

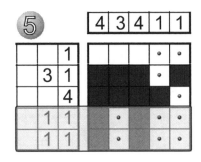

Column 5: Clue 1 is already on the grid. We can fill in the rest with background color.

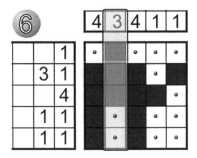

Row 3: There are only 4 squares left to place clue 4 on the grid.

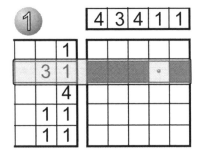

Column 4: Clue 1 is already on the grid. We can fill in the rest with background color.

Row 4/5: Clues (1,1), with 1 empty square between them, add up to 3 available squares.

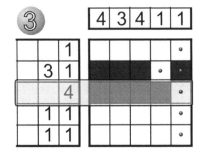

Column 2: There is only 1 square left to complete clue 3 and finish the puzzle.

Names

Nonograms are also known by many other names, including Paint by Numbers, Griddlers, Pic-a-Pix, Picross, PrismaPixels, Pixel Puzzles, Crucipixel, Edel, FigurePic, Hanjie, HeroGlyphix, Illust-Logic, Japanese Crosswords, Japanese Puzzles, Kare Karala!, Logic Art, Logic Square, Logicolor, Logik-Puzzles, Logimage, Oekaki Logic, Oekaki-Mate, Paint Logic, Picture Logic, Tsunamii, Paint by Sudoku and Binary Coloring Books.
(From Wikipedia, the free encyclopedia)

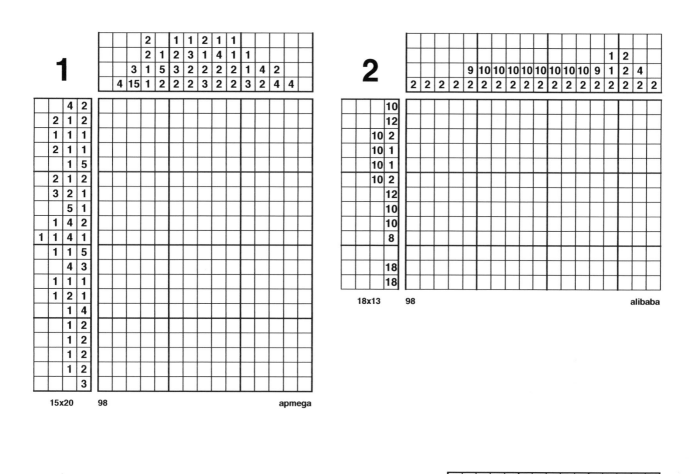

1

15x20　　98　　apmega

2

18x13　　98　　alibaba

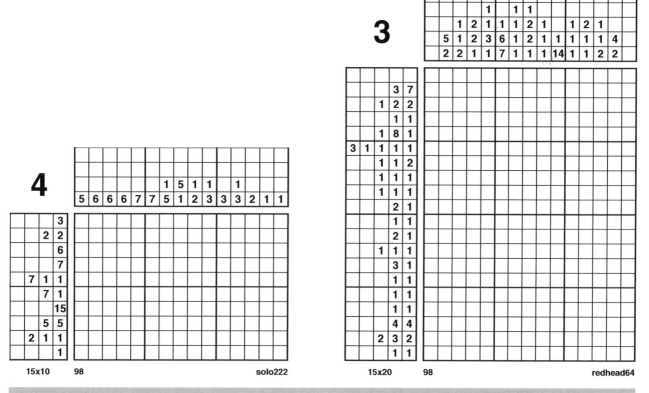

4

15x10　　98　　solo222

3

15x20　　98　　redhead64

5

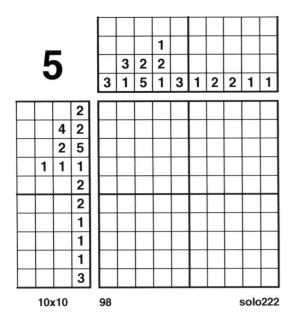

10x10   98   solo222

6

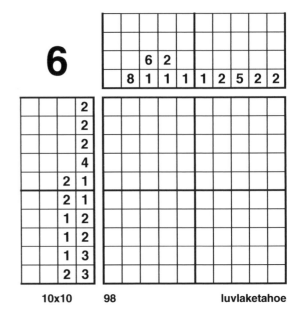

10x10   98   luvlaketahoe

7

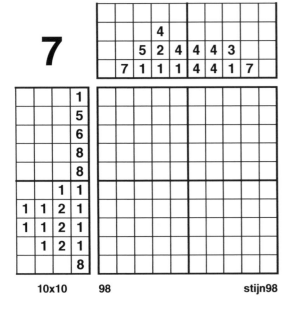

10x10   98   stijn98

8

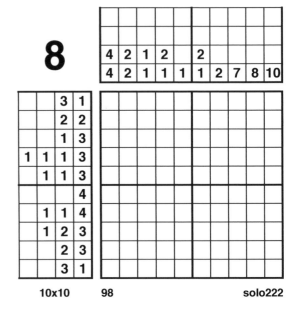

10x10   98   solo222

9

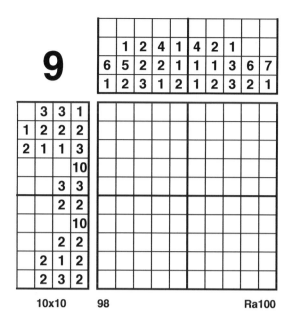

10x10   98   Ra100

10

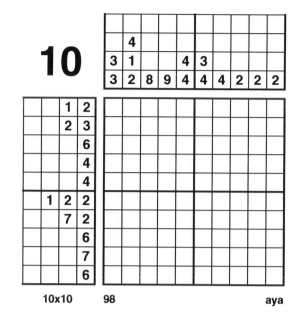

10x10   98   aya

11

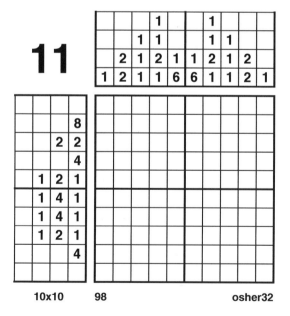

10x10   98   osher32

12

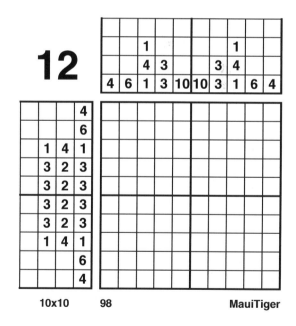

10x10   98   MauiTiger

13

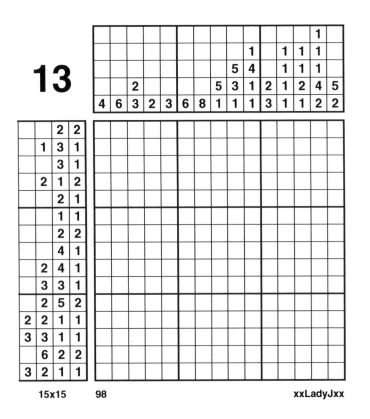

15x15   98   xxLadyJxx

14

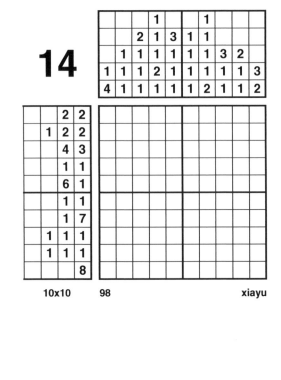

10x10   98   xiayu

16

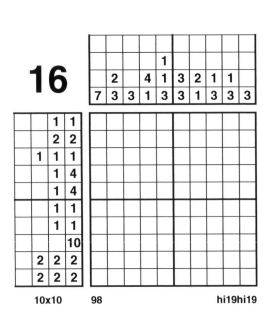

10x10   98   hi19hi19

15

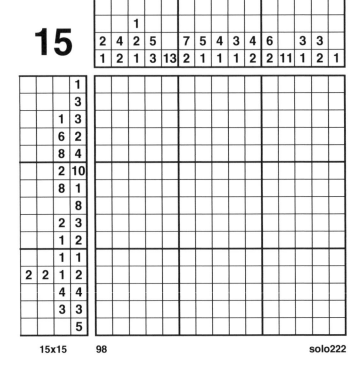

15x15   98   solo222

17

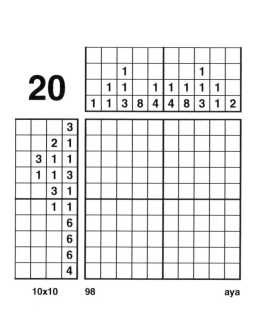

15x15   98   ferrari

18

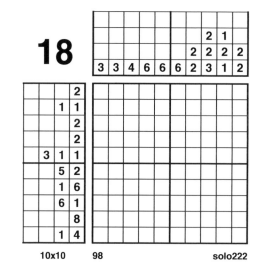

10x10   98   solo222

20

10x10   98   aya

19

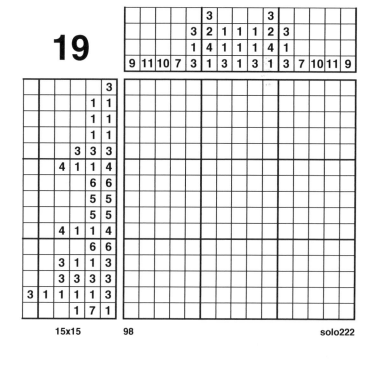

15x15   98   solo222

22

15x15   98   Rainbow15

21

15x15   98   tjbpm

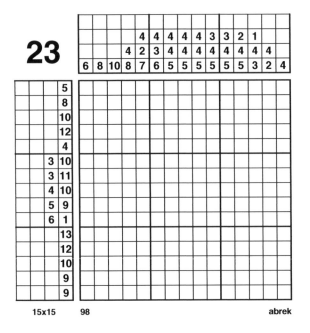

23

15x15   98   abrek

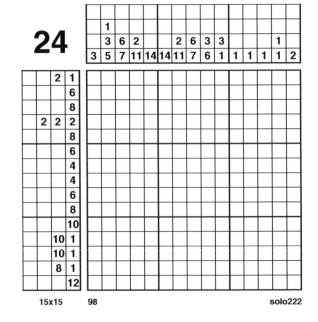

24

15x15   98   solo222

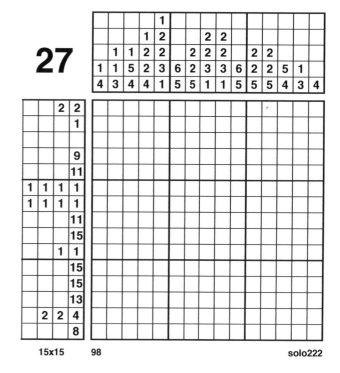

15x15　98　Gothic56

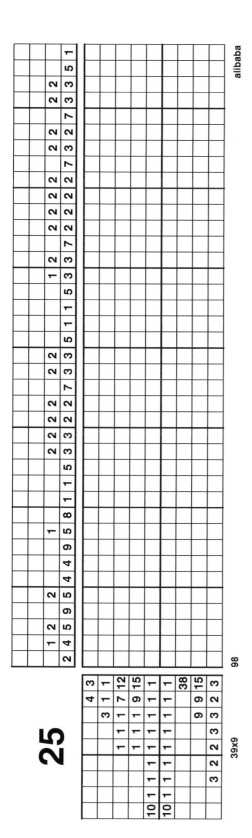

alibaba

98

25

39x9

27

15x15　98　solo222

28

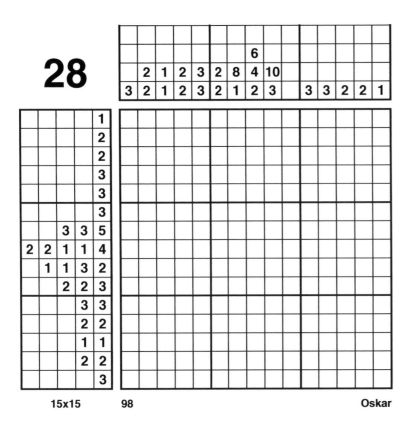

15x15   98   Oskar

29

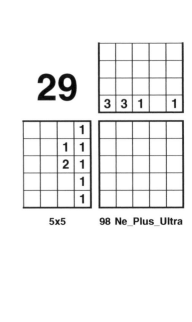

5x5   98 Ne_Plus_Ultra

31

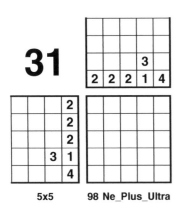

5x5   98 Ne_Plus_Ultra

30

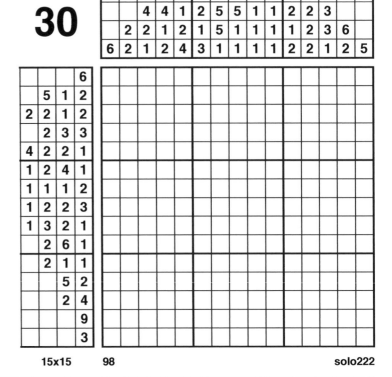

15x15   98   solo222

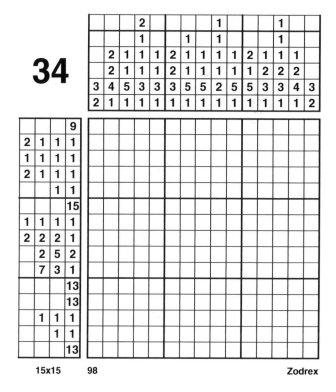

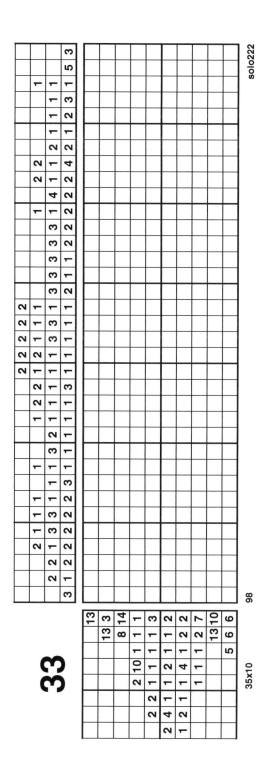

32

solo222

15x15   98

33

35x10

98

34

15x15   98

Zodrex

35

Column clues (top):

			2	5		1 5	2	6	2		5			3										
	6	5	4	9	10	2	10	6	3	9	6	9	2	8	6	6	2	4	3	3	1			
6	2	2	1	1	1	1	1	1	1	1	1	1	1	1	1	1	1	1	1	1	1	2	1	

Row clues (left):

				1	3
				1	5
			1	1	5
			1	5	3
				1	12
	1	2	4	6	
		1	9	6	
		1	20	1	
		1	18	1	
1	3	2	7	3	1
	1	8	5	1	
	2	12	2		
	1	6	1		
			3	3	
				6	

36

Column clues (top):

					1 1 1					1 1 1 1													
		2	2	1	1 1 1			1 1 1 1									1 1						
3	5	4	5	5	5	6	7	1	13	10	9	8	7	7	4	1	1		1	1			
1	2	1	1	1	1	1	1	12	1	1	1	1	1	1	1	2	1	1	1	1	1	4	1	1

Row clues (left):

		5
		6
		1
		3
	12	5
2	8	1
2	9	1
	2	17
		16
		15
		14
		11
		9
2	1	2
		16

37

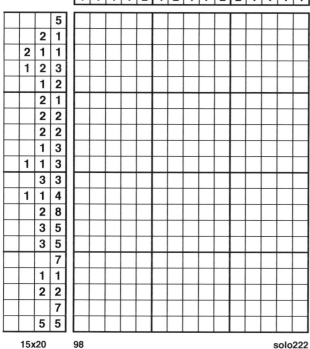

15x20   98   solo222

38

5x5   98 Ne_Plus_Ultra

39

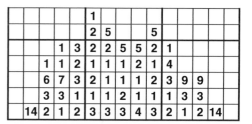

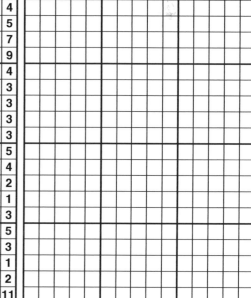

15x20   98   Oskar

40

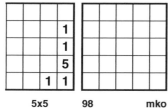

5x5   98   mko

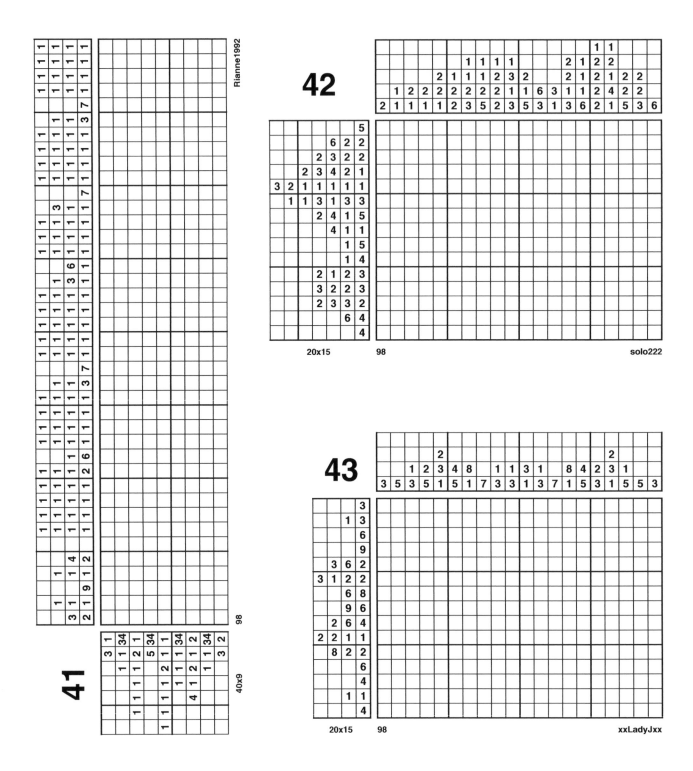

41

40x9

Rianne1992

98

42

20x15　　98

solo222

43

20x15　　98

xxLadyJxx

44

Row clues (15 columns of grid; 25 rows):

				1	1	1	3
			1	1	1	1	7
	1	3	1	1	1	4	2
2	1	1	1	1	5	1	1
						2	10
				4	13	3	
					16	3	
						20	
					16	3	
				4	13	3	
						2	10
2	1	1	1	1	5	1	1
	1	3	1	1	1	4	2
			1	1	1	1	6
				1	1	1	3

Column clues (bottom rows):

2												2			
1										2		1			
										1		2	1	2	3
4			2	1				1		1	3	3	1	1	1
1			3	5	7		1	7		1	5	3	1	1	1	3
4	9	5	5	3	2	1	11	1	14	7	14	7	13	1	9	9	2	2	3	2	2	2	3	3

25x15 98 MDE001

45

Column clues (top):

		1				1			
		4	1			1	4		
		2	1	1	1	2	2		
		1	3	1	3	1	3	1	
5	4	3	2	1	4	3	1	1	4	1	2	3	4	5
15	14	12	11	1	1	8	7	8	1	1	11	12	14	15

Row clues:

			15
	4	4	
	3	3	
	2	2	
1	5	1	
1	1	2	
1	2	1	
1	1	1	
			7
	1	1	
1	2	2	1
2	1	1	2
2	5	2	
3	5	3	
4	5	4	
4	5	4	
4	3	4	
5	3	5	
5	1	5	
	4	4	
	5	5	
	4	4	
	5	5	
	4	4	
	6	6	

15x25 98 nutrija

46

Column clues (top):

			2				
			1	1			
			1	1	1	10	
	1	5	1	1	1	1	6	1
7	2	1	1	11	1	1	1	1	1	3	8	1
1	2	4	4	8	13	8	8	8	1	14	8	4	4	2

Row clues:

			4
1	1	1	
2	1	2	
1	2	1	1
1	1	1	1
1	1	1	1	1
2	1	1	1	2
1	1	1	2	1	1
1	1	2	1	1	1
2	1	1	2	2
1	3	1	1	1
1	1	1	3
2	1	1	
4	1	1	
	1	2	
1	1	1	
	1	1	
			14
	8	5	
	7	4	
	7	4	
	5	2	
	5	2	
	5	2	
			8

15x25 98 vargben

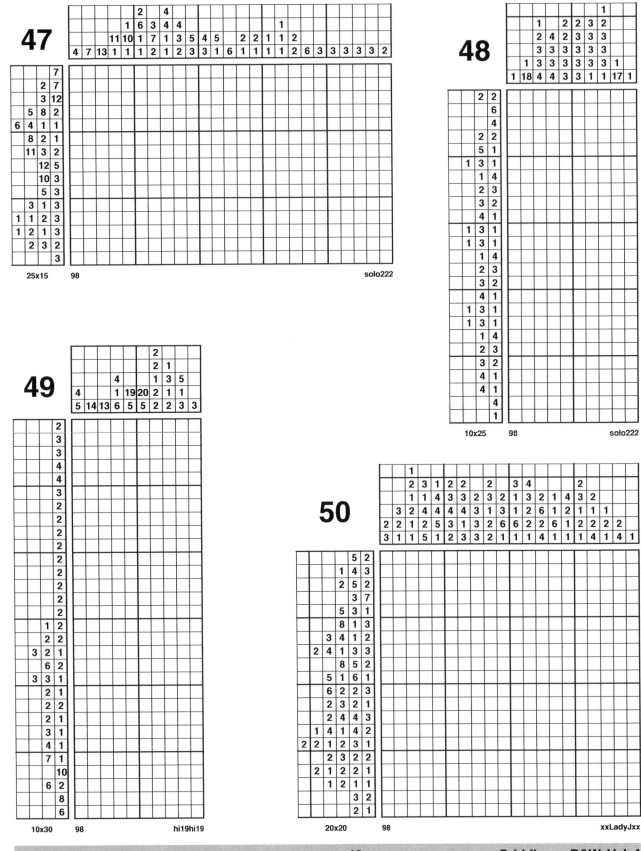

47

25x15   98   solo222

48

10x25   98   solo222

49

10x30   98   hi19hi19

50

20x20   98   xxLadyJxx

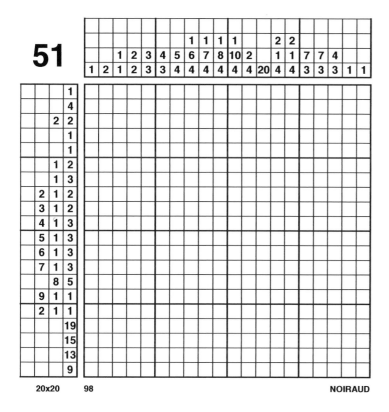

51

20x20   98   NOIRAUD

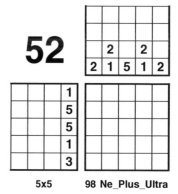

52

5x5   98 Ne_Plus_Ultra

53

20x20   98   nico5038

54

5x5   98 Ne_Plus_Ultra

55

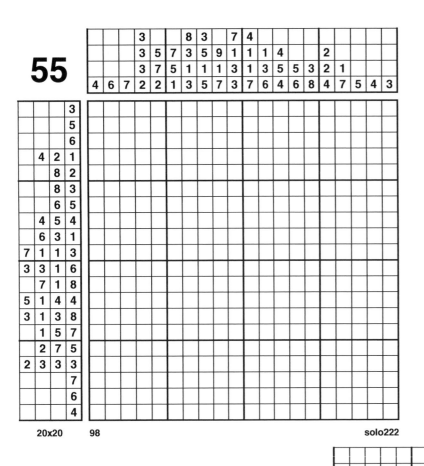

20x20   98   solo222

56

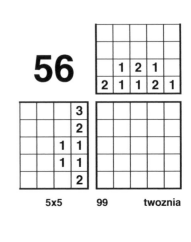

5x5   99   twoznia

57

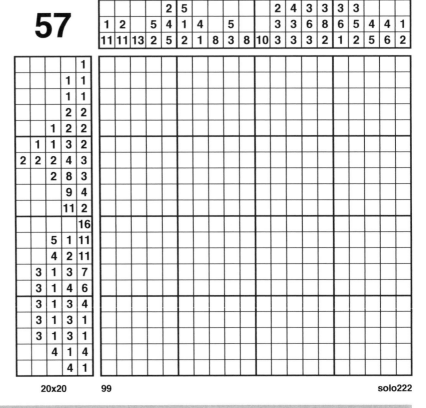

20x20   99   solo222

58

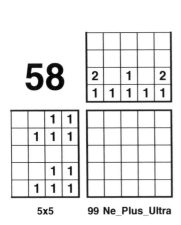

5x5   99 Ne_Plus_Ultra

59

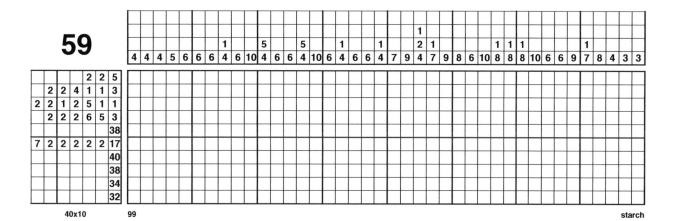

40x10   99   starch

60

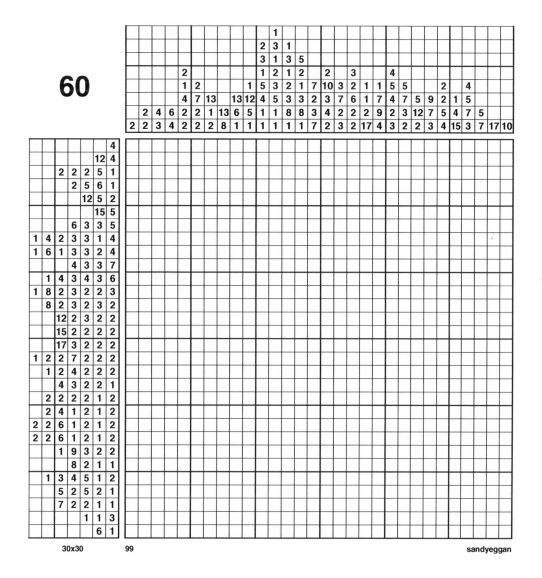

30x30   99   sandyeggan

61

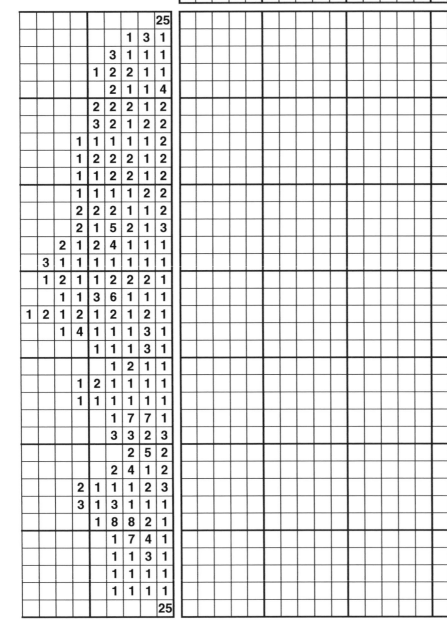

25x35   99   xiayu

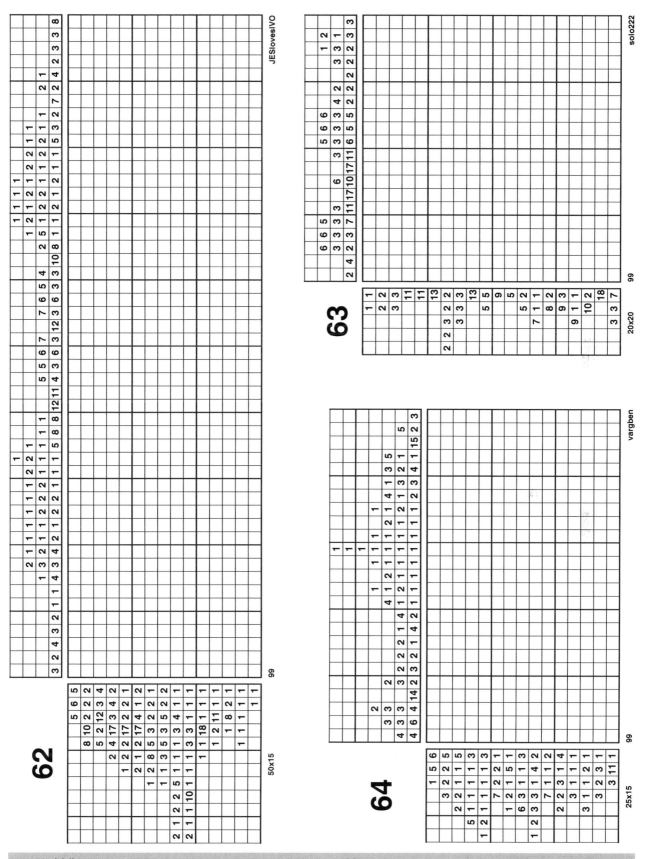

62

JESlovesIVO

50x15

99

63

solo222

20x20

99

64

vargben

25x15

99

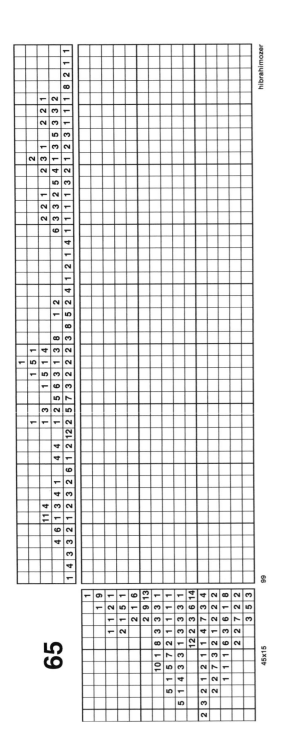

65

45x15

hibrahimozer

99

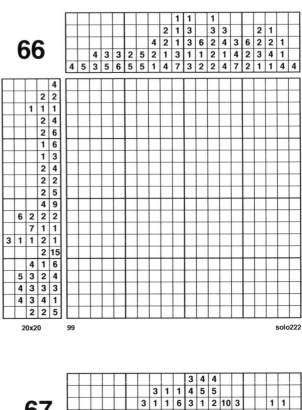

66

20x20   99   solo222

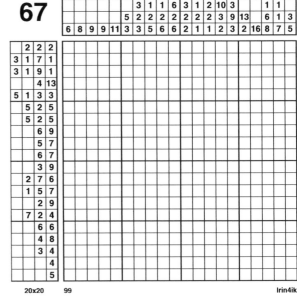

67

20x20   99   Irin4ik

68

elimaor

99

20x35

69

dromidror

99

20x20

70

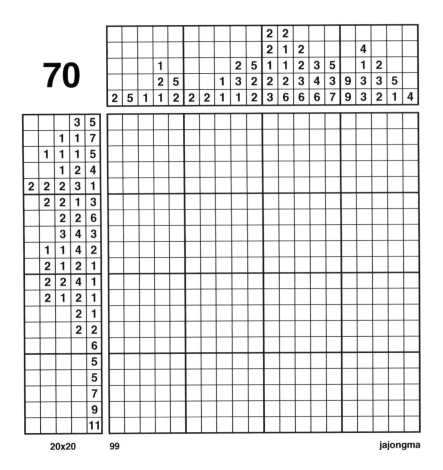

20x20   99   jajongma

71

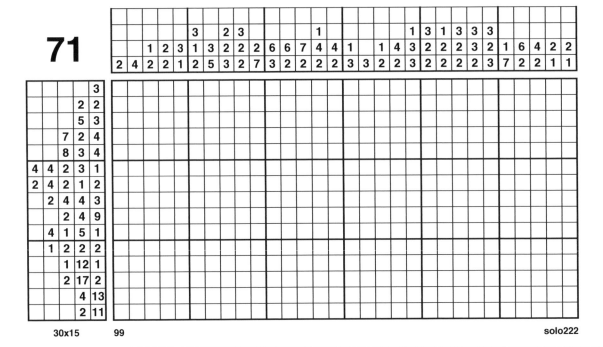

30x15   99   solo222

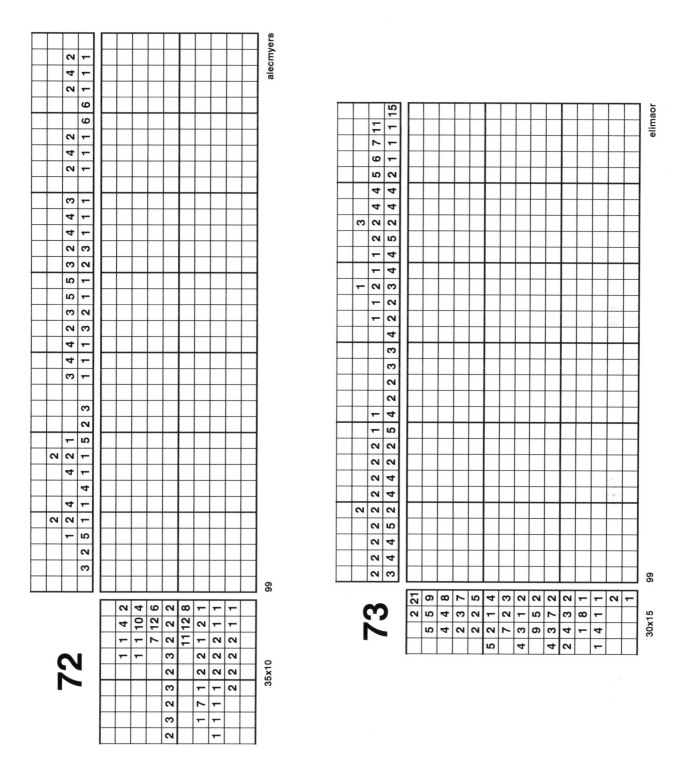

72

35x10

alecmyers

99

73

30x15

elimaor

99

74

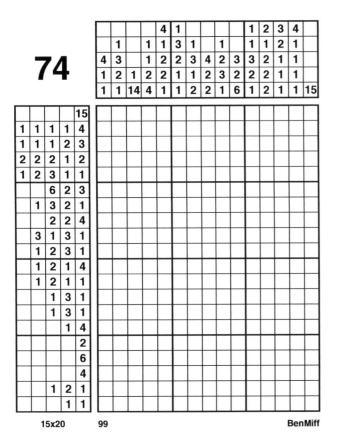

15x20   99   BenMiff

75

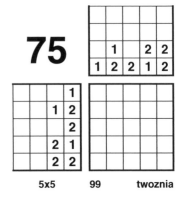

5x5   99   twoznia

76

15x20   99   airam

77

5x5   99   twoznia

78

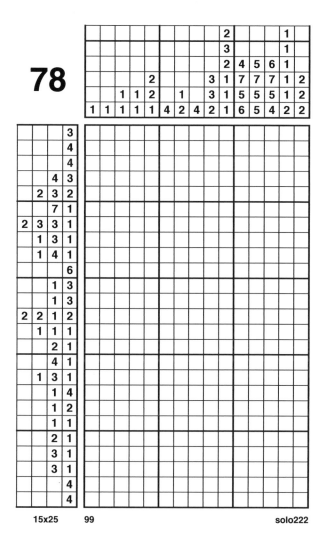

15x25   99   solo222

79

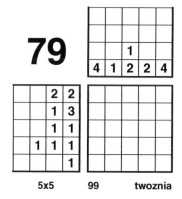

5x5   99   twoznia

80

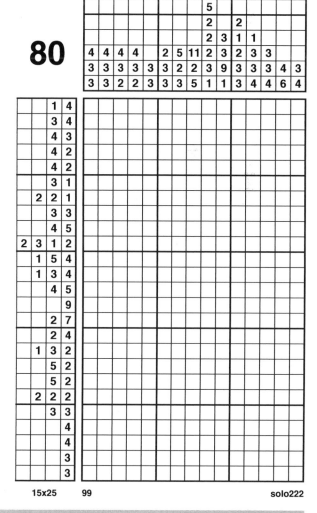

15x25   99   solo222

81

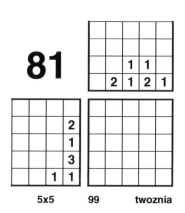

5x5   99   twoznia

82

20x15 — 99 — solo222

Column clues (top):

													2					
	1	1	1			5	6	6	8		2	2	5	8				
1	1	1	1	1	5	1	3	1	1	8	6	5	1	1	8	3	2	
4	1	3	1	1	1	1	3	1	1	1	3	1	3	1	1	1	1	4	1	3

Row clues (left):

			3
		1	1
1	1	1	4
	1	1	8
1	1	6	3
	4	7	2
		1	12
		2	11
		2	12
	1	7	2
	2	5	1
		3	4
	2	5	1
1	1	1	1
		5	5

83

20x15 — 99 — op1

Column clues (top):

			1						1			
	2	2	3					3		2		
2	1	2		2	2	2		2	2	2		2	1	2
2	2	1	7	1	5	4		4	5	1	7	4	2	2
4	6	3	5	3	2	2	2	2	9	9	2	2	2	2	2	3	5	3	6	4

Row clues (left):

					2	2
			4	2	2	4
			6	1	1	6
		2	2	2	2	2
2	1	3	2	3	1	2
				2	10	2
				5	6	5
						16
						11
				2	2	2
		2	1	2	1	2
				2	4	2
			3	2	2	3
					5	5
					3	3

84

25x20   99   nasa17

85

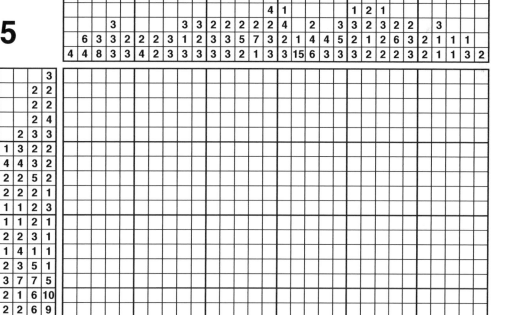

30x20   99   solo222

86

										1														
				1		3			2	4	4							4	1		4			
			1	2	4	3	2		2	1	2	4	1	1	1			4	1		1			
			2	1	1	1	3	5	1	1	1	3	2	2	3		5	2	3	6	1	4		
		5	1	3	1	2	1	2	1	1	3	2	3	5	8	18	10	2	1	3	1	4	1	
1	3	1	2	3	6	2	2	2	2	2	2	2	2	2	2	2	2	2	6	3	2	1	5	3

			1	1
				5
		1	3	3
	4	2	5	3
	3	4	3	3
	5	4	3	5
	5	2	1	3
2	2	1	2	1
	4	2	2	6
	2	2	4	3
			11	4
	1	2	5	2
			4	4
			3	4
			1	4
			3	5
			1	8
				17
			2	2
			2	2
			2	2
			2	2
			2	2
				19
				21

25x25 99 xiayu

87

25x25 99

Left (row) clues:

					6
					8
					8
	3	1		2	2
1	3	1	1	2	1
	1	2	2		1
		1	5		3
	3	1	1	1	2
	2	1	1	3	2
	3	1	1	2	1
	2	1	1	3	1
	3	1	1	2	2
	1	1	1		5
	1	1	1		1
	2	1	1		2
	3	1	1		3
	8	4	4		6
		2	3		2
		2	1		2
		2	1		2
		1	1		1
		8	1		8
	1	2	2		1
	1	1	1		1
			8		8

Top (column) clues:

C1	C2	C3	C4	C5	C6	C7	C8	C9	C10	C11	C12	C13	C14	C15	C16	C17	C18	C19	C20	C21	C22	C23	C24	C25
							1													1				
							1													1		1		
				5			1					3		3		4				1		5		
				1	12	1	2	2	5	3	1		1	3	1	5	1	1	12	1	2			
				2	3	4	1	2	1	11	1	3	1	11	1	2	1	4	3	2	1	2		
				1	1	2	1	1	1	1	2	2	1	2	2	1	1	1	1	2	1	1	2	4
1	1	1	1	3	2	1	1	1	1	1	1	3	5	3	1	1	1	1	1	2	3	1	1	

Rainbow15

88

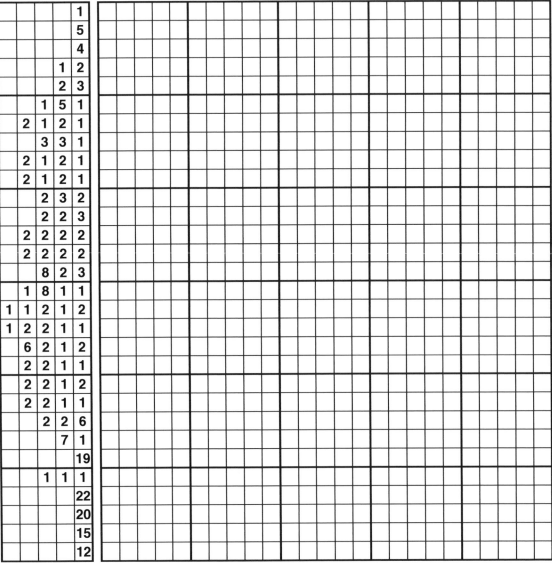

25x30   99   nasa17

89

30x30 — 100

Column clues (top):

				3		2		5	6	9						3	3						14						
	7	5	4	1		1	2	6	1	2	4		1			2	1	1			14		2						
	10	2	1	2	1	3	1	1	2	8	3	5	2	2	2	1	2	2	4	4	8	12	1	15	16	17	1		
16	3	1	2	2	1	2	1	3	3	2	1	1	5	5	8	4	5	2	1	1	1	3	2	1	1	1	2	2	15
10	7	7	5	2	2	2	1	1	1	1	2	1	2	3	8	11	11	12	10	9	9	7	8	5	4	4	1	1	8

Row clues (left):

				8	21
				15	13
			7	5	11
			5	4	9
			4	3	8
			3	2	8
		3	1	3	8
		2	2	2	8
		2	2	2	7
	3	1	3	1	7
	1	3	3	6	7
	1	2	4	5	7
	2	1	3	2	6
		5	1	4	6
2	3	2	3	3	1
		1	2	2	2
			2	2	2
		2	1	2	1
			1	1	1
			2	4	3
1	1	3	4	2	2
		1	4	7	1
		1	7	2	1
		3	7	9	1
	3	1	3	9	2
		4	3	11	3
			5	14	1
			6	14	2
	4	2	3	12	1
		4	6	13	1

syucesan

90

30x30

100

solo222

91

25x40 · 100 · elimaor

Column clues (left to right)

Col	Clues
1	5 2 2 5
2	2 3 3 3 5 9
3	4 4 1 4 1 4 4
4	2 5 3 2 3 4 3 4
5	2 3 4 3 2 4 2 6
6	2 4 2 4 2 4 4 6 3 7
7	3 3 2 4 1 3 3 1
8	5 2 4 5 1 4 4 4
9	5 5 7 1 1 4 5
10	2 1 2 6 1 3 5
11	1 5 4
12	4 1 1 1 5
13	11 3 1 4 4
14	3 2 3 6 6 3 2 6
15	2 4 2 7 4 1 6
16	1 6 1 2 9 5 1
17	1 5 2 3 2 8 7
18	6 1 5 3 3
19	1 6 6 6 2
20	1 2 3 3 5 7 8
21	3 3 3 3 3
22	4 4 3 4
23	7 8 4 9
24	2 2 4 4
25	6 3 2 2

Row clues (top to bottom)

Row	Clues
1	4 1
2	8 5 2
3	2 4 3 1 2
4	4 3 3 3 1 1
5	5 2 3 3 1 1
6	8 2 4 3
7	7 2 4 1
8	4 2 5 1
9	2 4 1 1
10	4 1 1 1 1
11	11 1 3
12	4 1 1 3 3
13	4 5 2 3 2
14	3 6 1 2 1
15	2 6 2 3
16	8 1 2
17	1 1
18	1 3 3 3 2
19	5 9 1
20	6 4 4 1
21	2 4 3 5 1
22	4 3 3 4
23	5 2 3 2 4
24	5 2 2 2 6
25	4 1 3 6
26	4 1 1 1 4 5
27	7 1 1 1 3 1 1
28	7 1 1 5 5
29	5 1 5 6
30	2 2 2 4 6
31	5 3 3 3 1 2
32	5 8 1 5
33	4 1 8 6
34	3 3 5 7
35	4 4 2 1
36	4 3 5 3
37	3 4 6 2
38	8 9
39	2 4 3 4
40	3 3

92

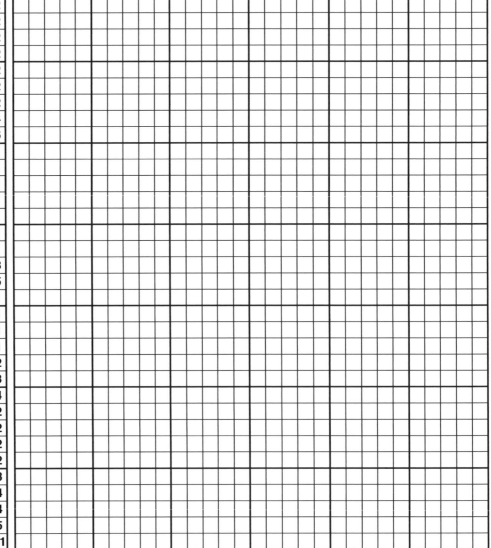

30x35

100

93

Id5

100

45x25

94

elimaor

100

40x30

95

35x35   100

eleonor

96

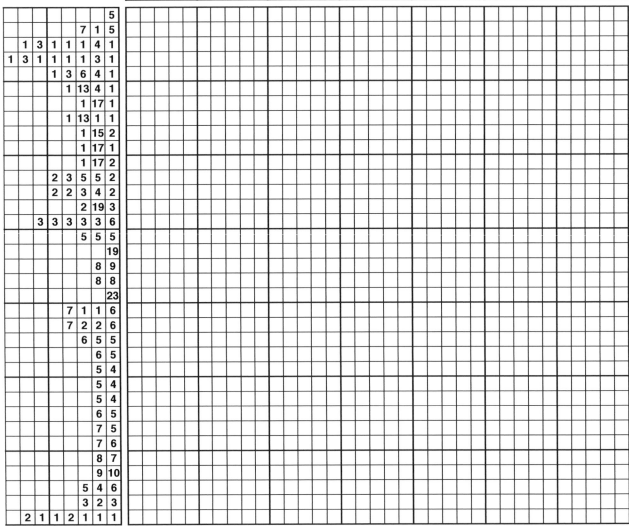

35x35   100

poj

ylletrollets

100

40x35

98

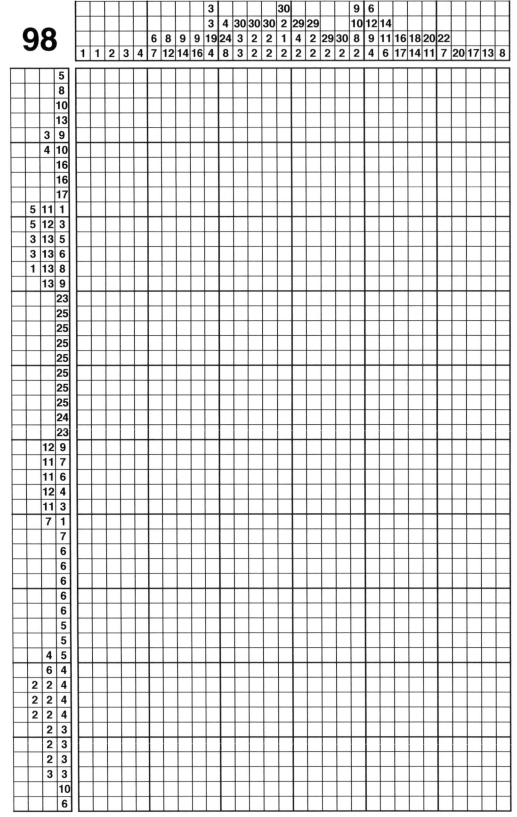

30x50   100

coachkelly

99

100

40x40

100

poj

100

45x35

101

						1		10																															
					3	5	8	2	14								3	3									5			7						6			
			5	3	2	2	2	2	2		6		6	5	4	4	2	2	3	3	3	3	4	5	6	5	5	5	6	5	1	2	3		5	5	7		
			4	5	3	4	3	5	3	18	10	5	12	13	19	19	14	13	2	2	3	2	2	3	2	3	2	4	5	3	14	13	12	4	7	3	5	9	
	9	7	6	5	3	1	3	5	1	3	2	12	6	6	2	3	4	5	14	3	2	2	2	2	3	4	4	4	4	4	8	7	6	10	1	7	4	7	
40	16	14	5	3	2	2	1	1	7	9	11	12	5	4	3	3	2	1	1	12	11	13	9	9	8	4	6	8	5	1	2	3	3	10	8	1	2	3	40

			7	13	9
			6	15	8
			6	17	7
		4	7	6	6
		4	6	6	5
		3	6	6	4
		3	5	4	3
		2	6	4	2
		2	5	4	2
		1	5	5	1
	1	5	6	5	1
	1	6	8	5	1
	1	10	2	5	1
	1	10	1	6	1
1	10	5	1	6	1
		1	17	10	1
	1	17	2	10	1
		1	17	10	1
1	2	11	4	8	1
	1	2	20	6	1
1	2	17	3	3	2
	1	3	11	2	3
		1	2	15	4
		1	2	22	5
		2	4	18	4
		2	3	18	3
		3	2	20	2
	3	2	14	3	1
	5	6	7	3	1
	6	6	6	5	1
	7	6	4	8	1
		6	10	9	1
5	5	4	4	5	1
4	5	3	4	4	1
3	5	2	4	4	1
4	6	1	3	4	1
4	6	1	3	4	1
	5	7	2	4	2
	7	9	1	4	3
		9	5	3	4

40x40

100

amam100

102

40x40

100

elimaor

103

40x40

101

elimaor

104

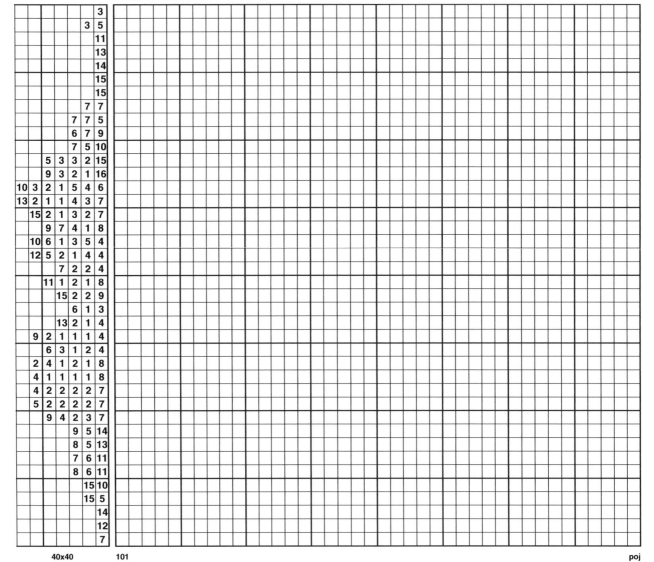

40x40   101

poj

105

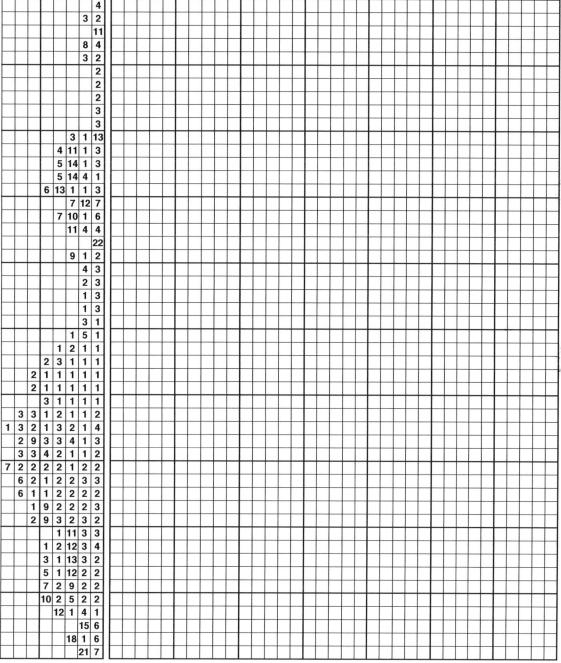

35x50   101   elimaor

106

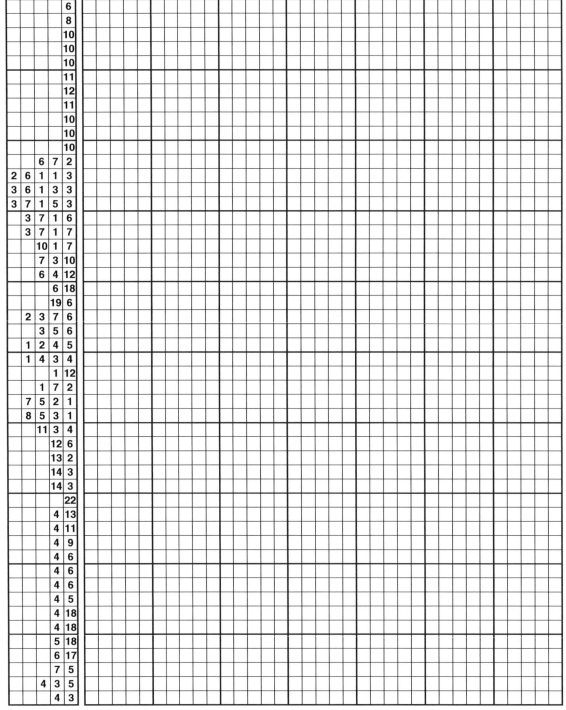

elimaor

107

101

50x35

108

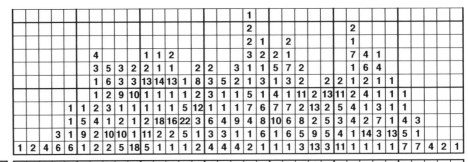

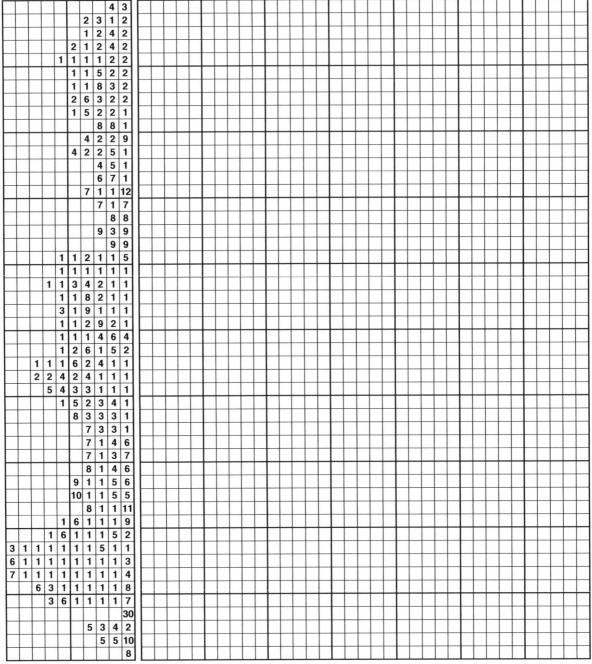

109

jfred99

101

45x40

110

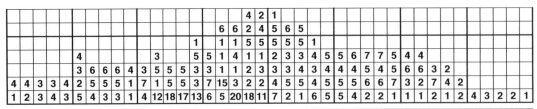

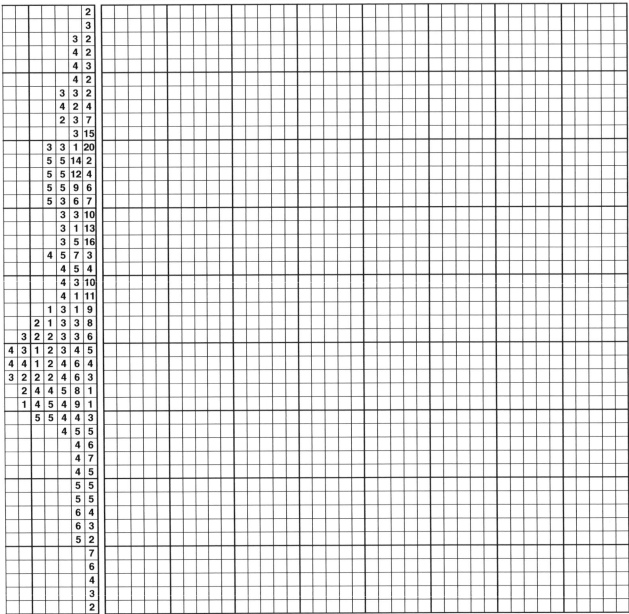

40x45   101

elimaor

111

49x38

101

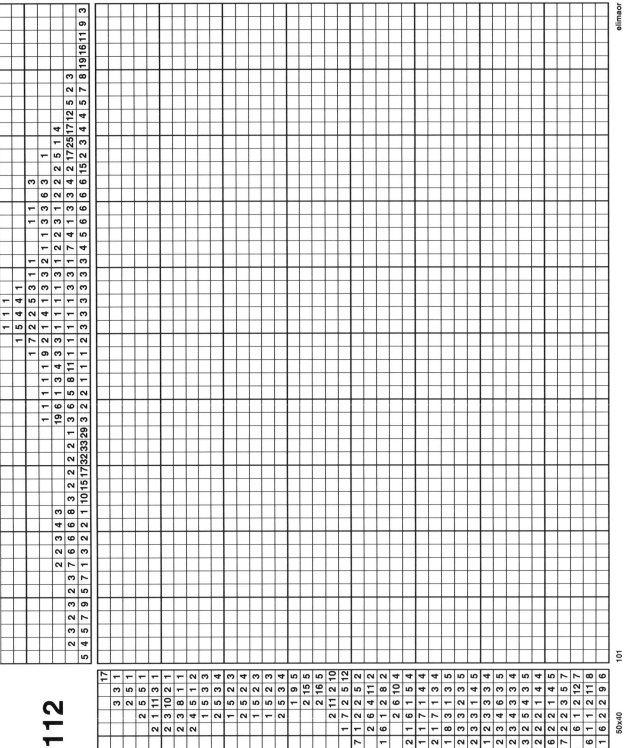

112

113

40x50   101

Column clues (top)

1	2	3	4	5	6	7	8	9	10	11	12	13	14	15	16	17	18	19	20	21	22	23	24	25	26	27	28	29	30	31	32	33	34	35	36	37	38	39	40
			2																																				
		2	2	3	3																																		
	2	2	3	2	2	3	3	1			4																												
1	1	2	7	2	2	4	5	5	4	4	6	8													5	5	5	5	6	6									
3	2	4	8	7	2	6	7	8	9	9	10	10	3	2	7	5						26	23	16	6	6	12	21	6	6	6	6	6	4					
3	3	2	3	4	2	1	2	8	10	11	13	14	28	28	29	30	32	33	34	31	27	1	1	3	4	5	8	13	35	34	32	31	29	26	24	21	18	14	10

Row clues (left)

1. 2
2. 4
3. 6
4. 8
5. 10
6. 2 10
7. 4 5 4
8. 5 3 4
9. 5 2 3
10. 4 5 1 2
11. 6 5 1 1
12. 3 4 3 2
13. 2 1 3
14. 2 2 1 5
15. 1 4 1 6
16. 1 4 6 3 7
17. 7 3 8
18. 7 4 8
19. 11 6 9
20. 3 3 7 9
21. 9 9
22. 4 11 10
23. 4 13 10
24. 5 16 10
25. 4 18 10
26. 22 11
27. 22 11
28. 22 12
29. 22 12
30. 22 12
31. 21 12
32. 17 12
33. 13 11
34. 13 11
35. 14 12
36. 14 12
37. 16 12
38. 17 11
39. 17 11
40. 17 10
41. 17 11
42. 2 17 10
43. 4 16 9
44. 3 15 9
45. 3 14 8
46. 2 12 8
47. 10 7
48. 8 4
49. 5 3
50. 1

oko

114

Glucklich

101

50x40

115

50x40

102

116

117

Amit11

102

45x45

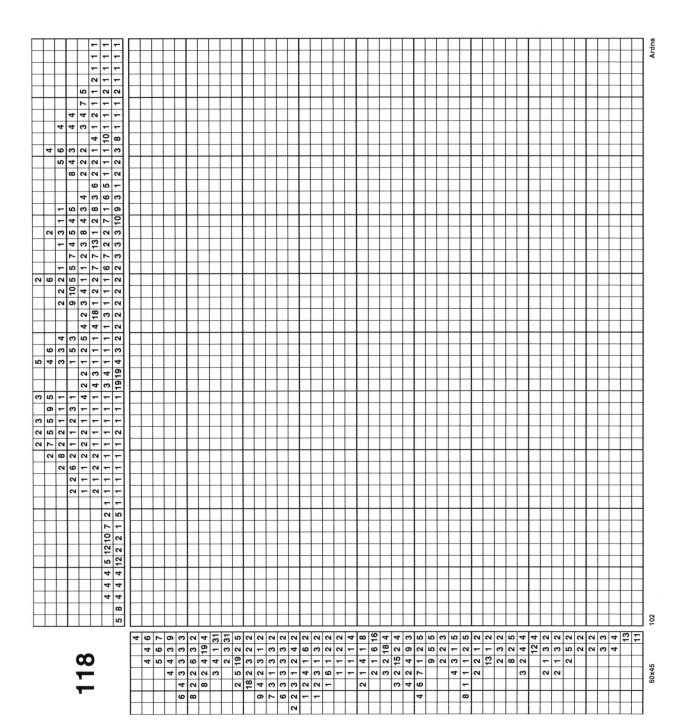

118

119

50x45

elimaor

102

120

45x50   102

arcadedweller

121

45x50

102

elimaor

122

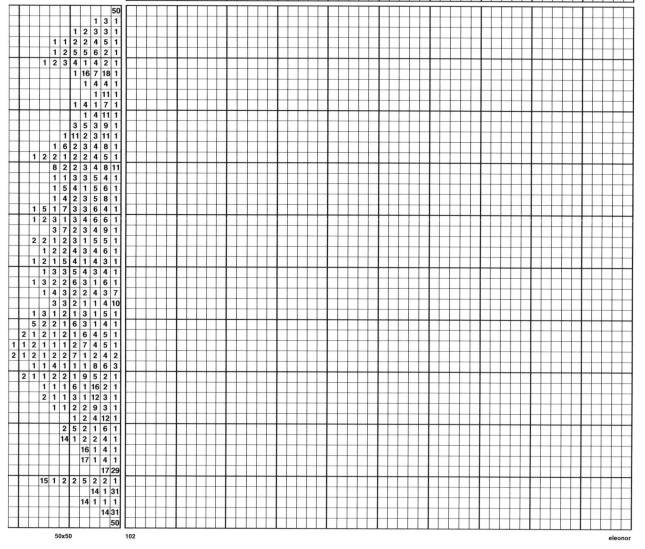

50x50   102   eleonor

123

50x50

124

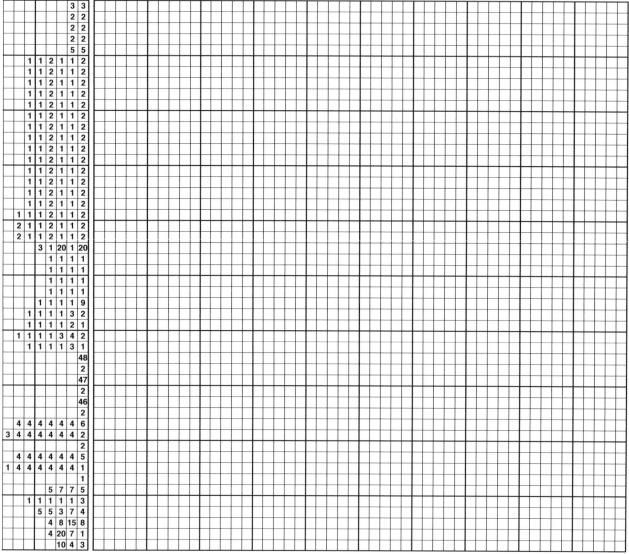

50x50   103   hi19hi19

125

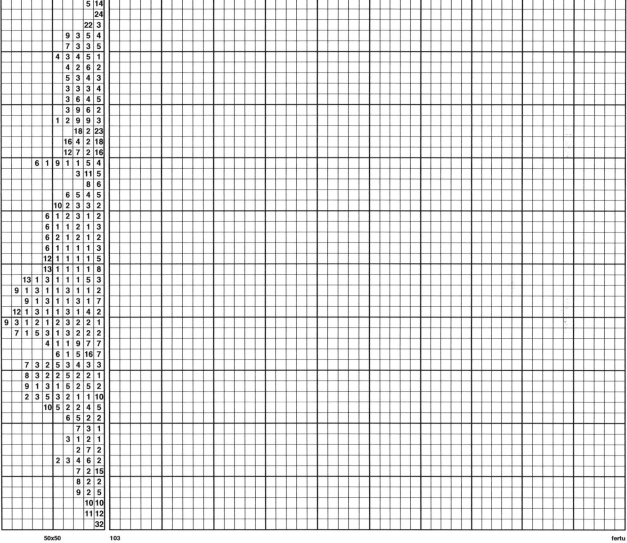

50x50   103

fertu

126

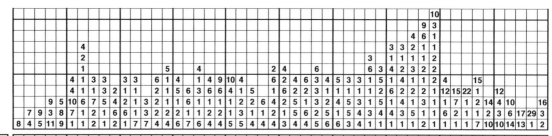

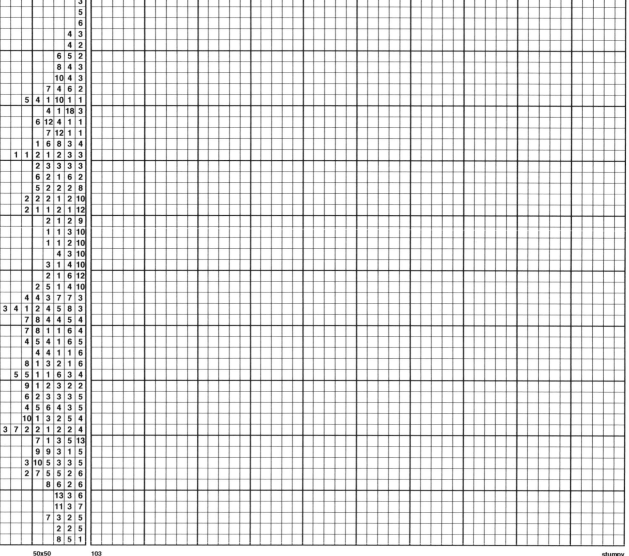

50x50   103

stumpy

127

50x50

128

50x50   103   elimaor

129

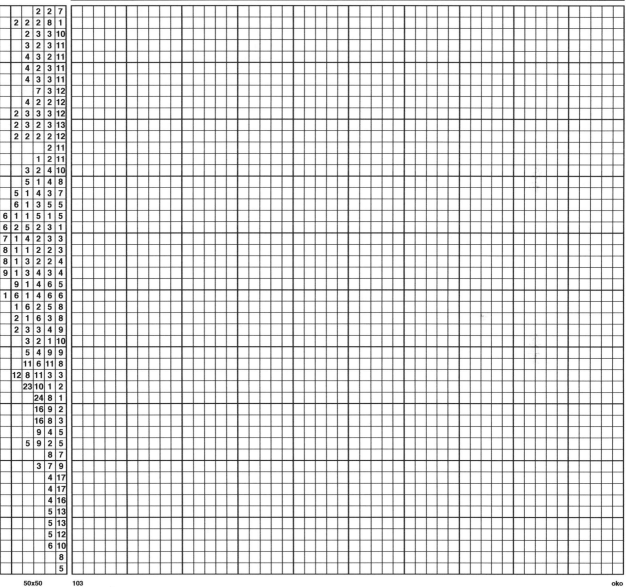

50x50   103

oko

130

50x50   103

eleonor

131

30x50 103

griddlers_books

132

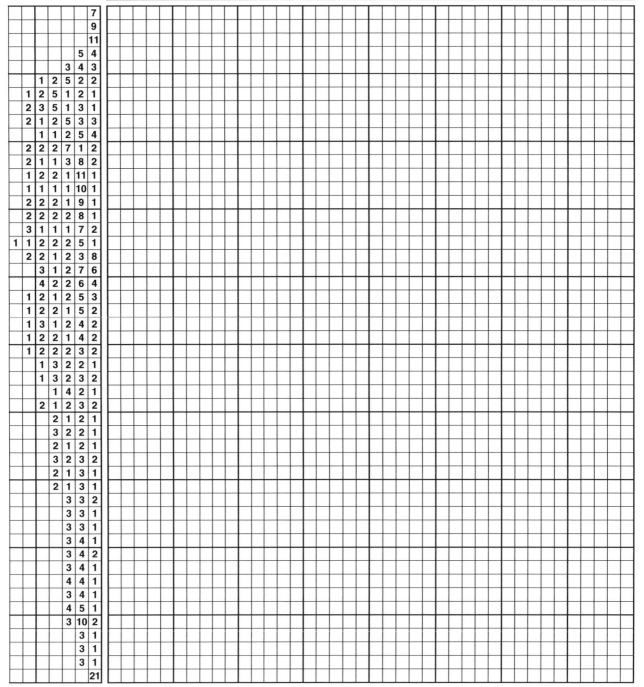

133

Size: 40x50 — No. 104

Column clues (top, left→right)

```
                                  4
                              1  8     5
                         3         2  6  3  2
                         5      2 16  1  2  4  8      20 13 13 13 10      2      4
                  2      6  1 15 15  1  3  7 17  6 18  6  4  2  6  2  9  5  7  8  2  3  1
      4  4  4  2  4      1  6  7  3  3  1 12  4 19 14  5  8 10  5  7  8  8  1  2  2  2  2  1  2  1  3  1  3  3  4
      6  3  2  2  1  1  6  7  3  1  2  4  8  3  8  3  3  2  2  2  2  2  2  3  3  4  6  7 13 10 13 11 10 10  9  9  9 10 12  9
```

Row clues (left, top→bottom)

Row	Clues
1	4
2	5
3	8
4	10
5	11
6	11
7	10
8	11
9	4 8
10	1 11
11	1 1 7
12	2 6
13	1 6
14	4
15	5
16	7
17	2 6
18	5 6
19	7 3 1
20	8 1 3
21	9 4
22	12
23	12
24	3 7
25	3 6 7
26	3 17
27	3 15 3
28	8 9 1
29	10 7 3
30	13 6 1
31	14 5 2
32	15 2 2 1
33	6 6 1 1 2
34	4 5 1 1 1
35	1 2 3 1 1 1
36	2 4 3 1 1 1 2
37	1 2 1 1 1 1 1
38	1 3 2 1 1 1 2
39	2 6 1 1 1 3
40	1 3 2 1 1 2 4
41	1 3 1 1 12
42	2 2 1 1 12
43	1 1 1 1 12
44	1 1 1 13
45	2 1 1 14
46	1 1 1 14
47	1 2 1 14
48	2 6 15
49	4 26
50	6 25

40x50   104

134

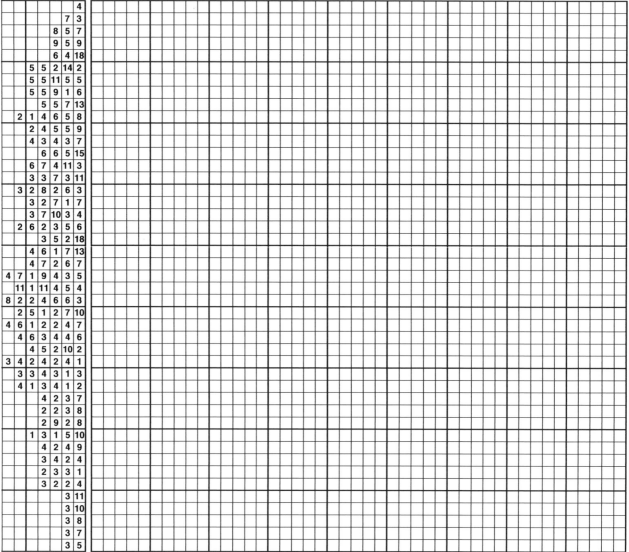

135

																										1	1																						
																									1	2	1	1																					
													1	1	1									1	2	2	1	2	1	1	1	1	1	1	1													
										1	1	1	2	2	2	1		1	1	1	1	1	1	2	2	2	1	2	2	2	2	2	2	2	3	1	1	1	1	1	1								
				1	1	1	1	1	1	1	2	2	2	2	2	2	1	2	2	2	2	2	2	2	1	1	2	1	2	1	2	2	3	2	3	2	3	2	3	2	3	1	1	1	1				
	1	2	3	2	2	2	2	2	2	1	2	2	1	1	1	2	2	2	1	2	2	2	2	1	2	4	5	1	2	1	2	2	1	2	2	2	3	3	4	3	3	2	3	4	4	1			
	2	3	2	2	2	2	2	2	2	5	6	7	6	8	10	1	2	1	1	1	1	1	2	3	11	6	5	11	11	9	6	1	2	2	4	2	2	2	2	2	2	3	3	3	4	4	1	1	
	1	1	2	3	4	4	5	6	7	3	2	3	3	3	3	12	2	12	10	8	6	5	3	10	1	1	2	3	3	5	5	7	8	10	12	22	21	20	18	16	14	3	3	3	4	4	4	4	
	3	4	4	3	2	2	3	3	2	2	2	2	2	1	2	2	18	3	3	2	2	6	11	3	2	2	1	2	1	2	1	2	1	2	1	1	1	1	1	1	1	13	10	4	5	9	4	4	
50	1	1	1	1	1	1	1	1	1	1	1	1	1	1	1	3	2	1	1	1	1	1	1	1	1	1	1	1	1	1	1	1	1	1	1	1	1	1	1	1	1	2	2	4	1	1	1	1	50

							50
					1	2	1
					2	3	1
					3	6	1
				1	2	14	1
				1	4	16	1
				1	6	12	1
				1	13	8	1
				1	16	6	1
				2	10	5	1
			1	2	4	5	1
			1	15	3	6	1
		1	9	7	4	5	1
		1	13	4	5	4	1
		1	16	4	5	4	1
			1	16	3	4	5
		1	2	15	2	4	4
		1	2	16	3	4	3
	1	1	12	4	2	3	2
	1	2	9	3	2	3	1
	1	2	8	6	2	3	1
1	1	6	5	2	2	3	1
1	2	5	5	2	2	3	1
	1	1	4	6	2	2	5
	1	2	3	6	3	2	4
	2	1	2	6	4	2	3
2	2	1	2	3	3	1	2
3	1	1	2	2	4	2	1
1	1	3	2	3	4	1	1
	1	1	2	6	5	2	1
	1	2	1	6	5	1	1
	1	1	2	5	6	2	1
	1	1	1	5	7	1	1
	1	2	2	5	8	1	1
	1	1	1	5	9	2	1
	1	2	2	4	10	1	1
		1	2	5	12	1	1
		1	2	3	14	1	1
			1	1	23	1	1
		1	2	2	18	1	1
		1	1	2	17	1	1
		1	2	2	15	1	1
		1	2	3	12	2	1
		1	2	3	10	1	1
		1	2	3	8	1	1
		1	2	3	2	1	1
			1	3	7	2	1
			1	2	1	1	1
				1	2	3	1
							50

50x50

104

griddlers_books

136

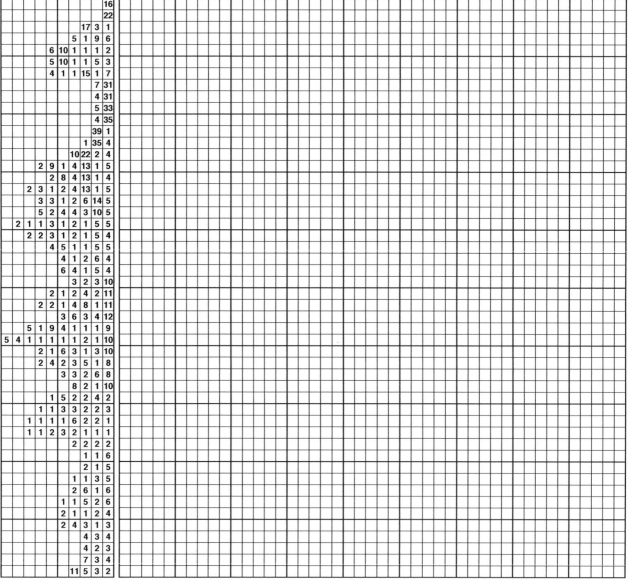

45x50   104

137

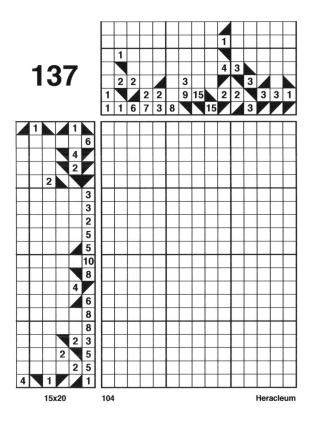

15x20 104 Heracleum

138

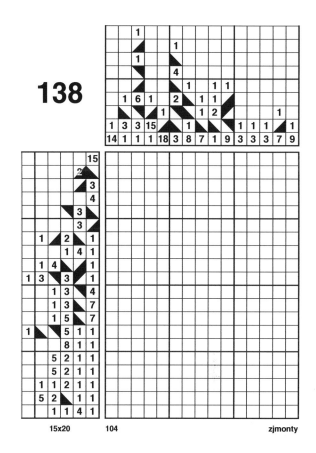

15x20 104 zjmonty

139

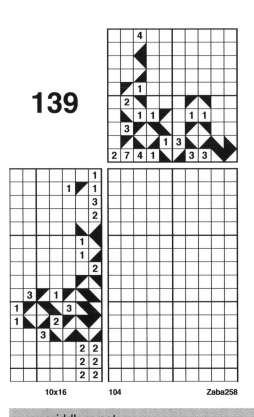

10x16 104 Zaba258

140

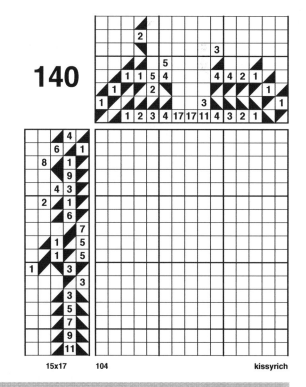

15x17 104 kissyrich

141

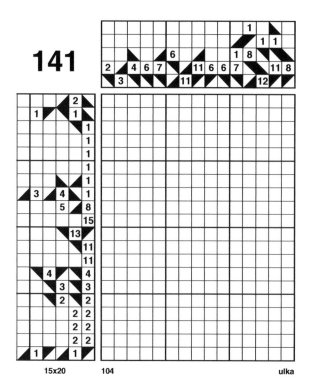

15x20   104   ulka

142

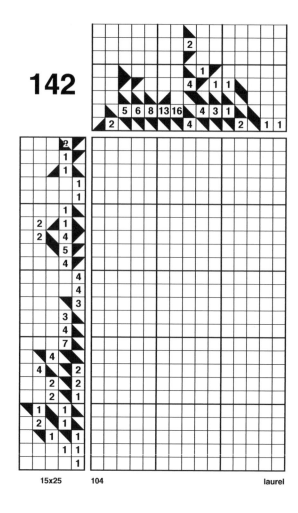

15x25   104   laurel

143

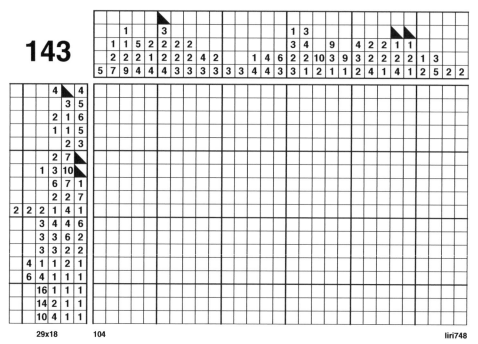

29x18   104   liri748

144

29x19   104   liri748

145

30x19   104   liri748

146

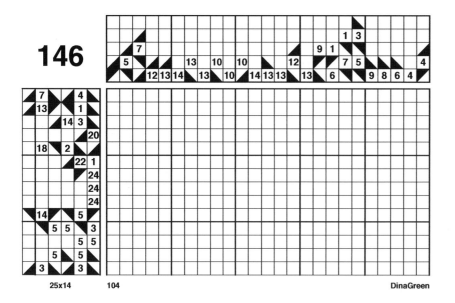

25x14   104   DinaGreen

147

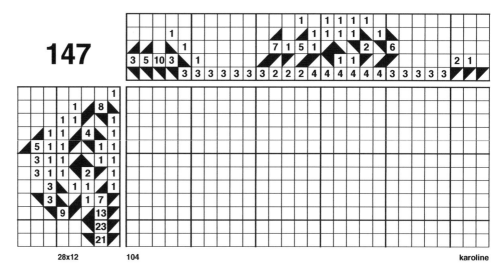

28x12   104   karoline

148

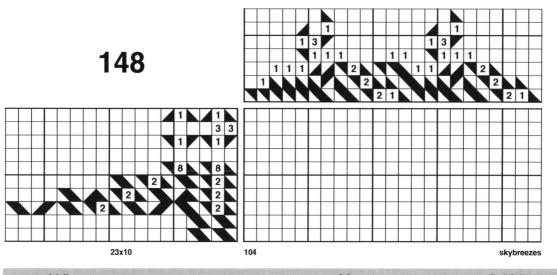

23x10   104   skybreezes

149

25x30   104   LuB

150

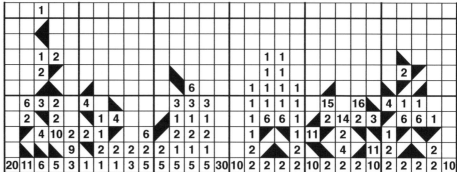

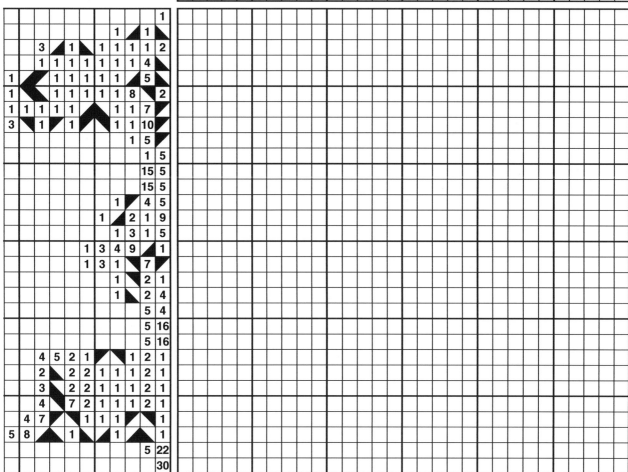

30x30   104

HSpring

151

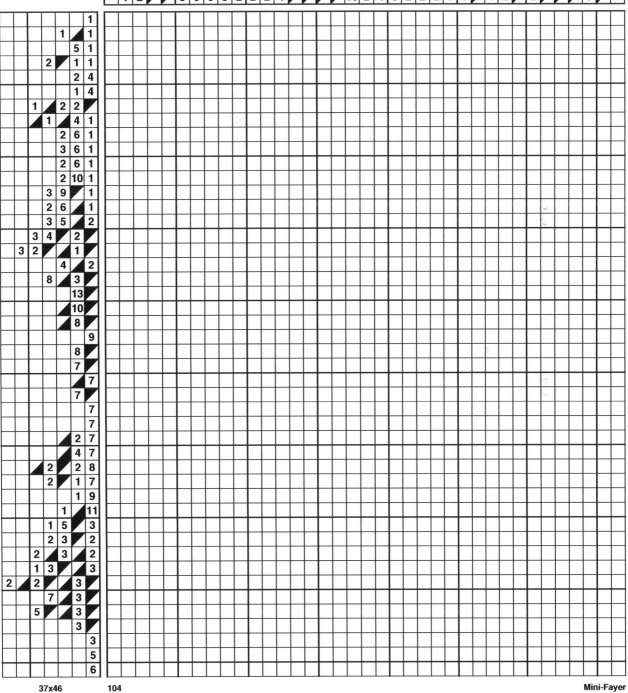

37x46   104

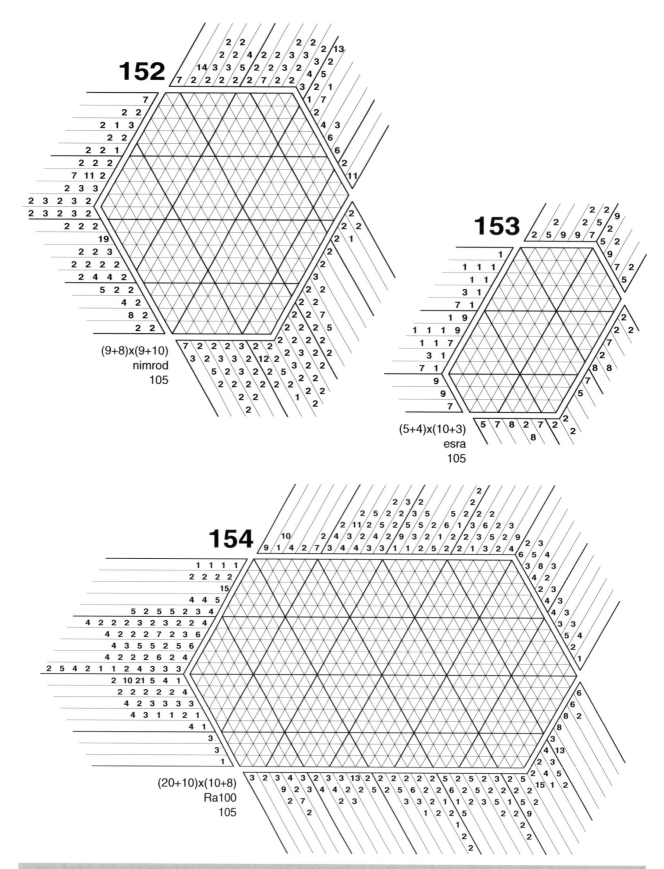

152

(9+8)x(9+10)
nimrod
105

153

(5+4)x(10+3)
esra
105

154

(20+10)x(10+8)
Ra100
105

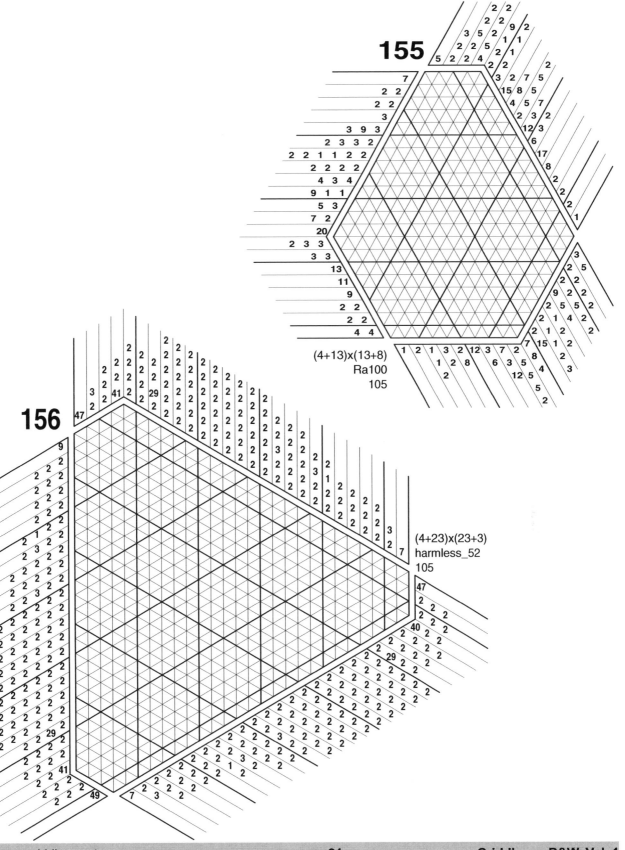

155

(4+13)x(13+8)
Ra100
105

156

(4+23)x(23+3)
harmless_52
105

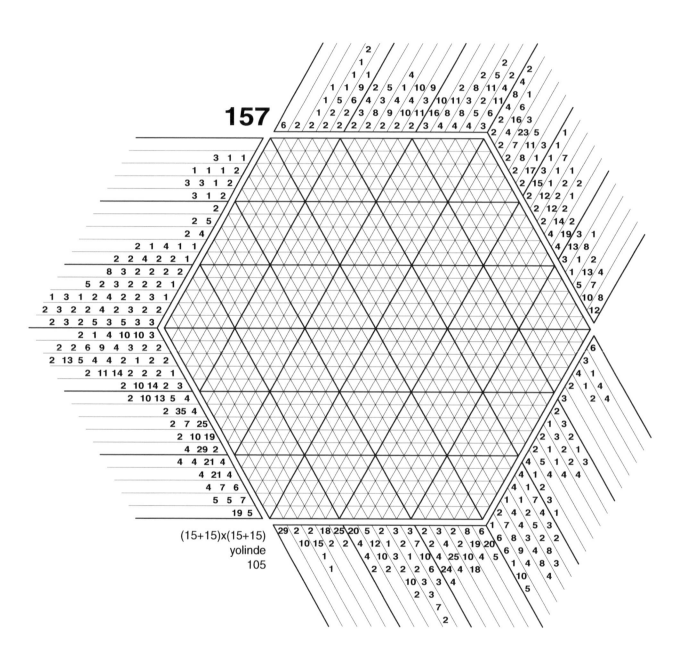

157

(15+15)x(15+15)
yolinde
105

158

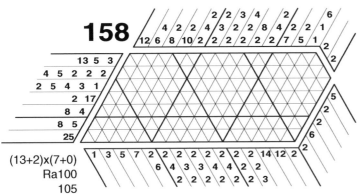

(13+2)x(7+0)
Ra100
105

159

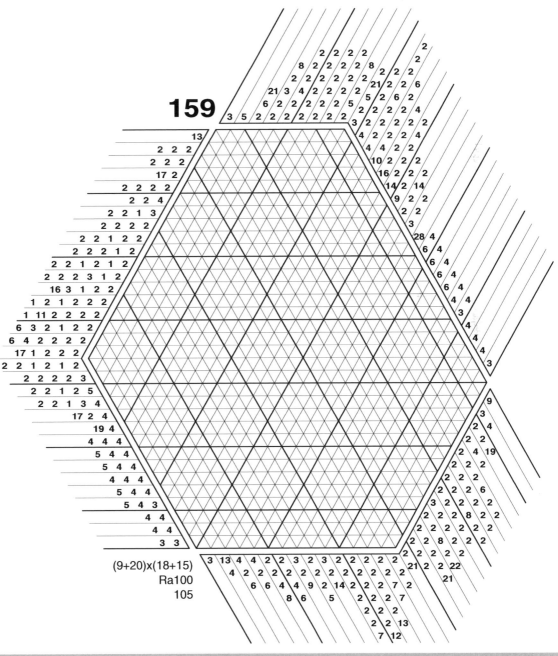

(9+20)x(18+15)
Ra100
105

160

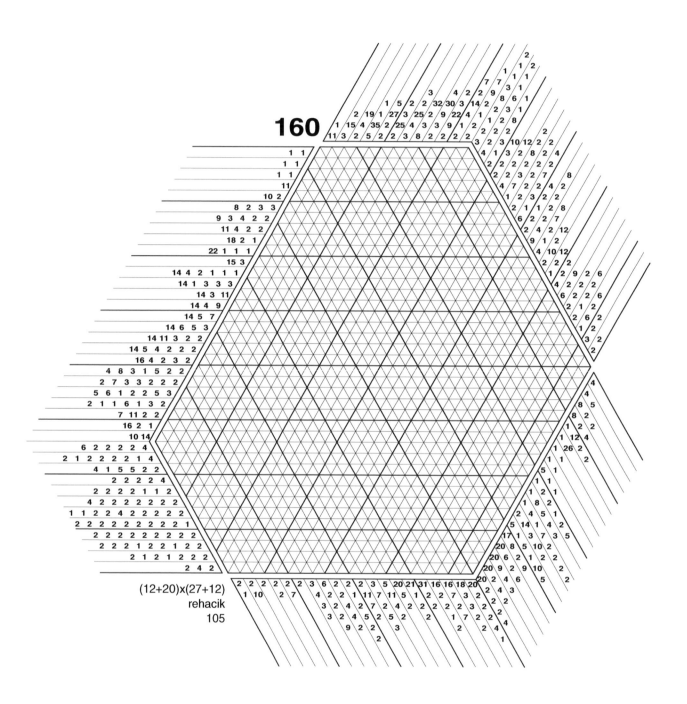

(12+20)x(27+12)
rehacik
105

161

(22+22)x(22+22)
rehaclk
105

162

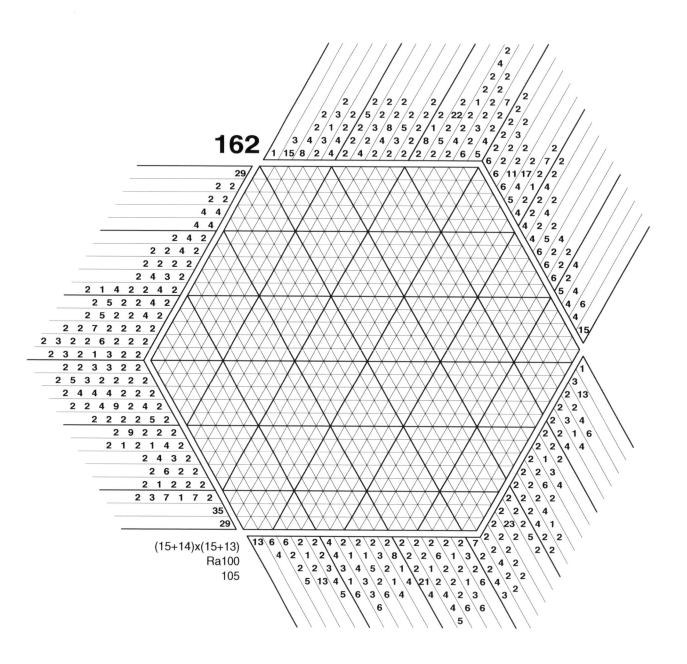

(15+14)x(15+13)
Ra100
105

163

(11+10)x(23+14)
Ra100
105

Solutions

 1: Ice Cream

 2: Coffee

 4: Peeping from the Nest

 5: Palm Tree

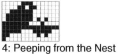 6: Time for Your Treat

 3: Hammer Head

 7: Small House

 8: I See the Moon... The Moon Sees Me...

 9: Sunrise

 10: Bunny

 11: Eye

 12: Tennis Ball

 13: Cat and...

 14: Snake

 16: Skate

 15: Rocking Horse

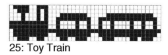 17: Fruits

 18: Little Fish

 20: Cactus

 19: Angel

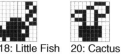 22: Well

 21: Old Diskette

 23: Helmet

 24: Watching

 25: Toy Train

 26: Chicken

 27: Cruise Ship

 28: Scissors

 29: Baseball

 31: Kite

 30: Rose

 33: Racing Car

 32: Bell

 34: One Tasty Burger

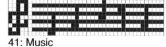 35: Watermelon

 36: Helicopter

 37: Bird

 38: Playing the Piano

 40: Table Tennis

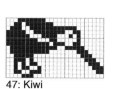 41: Music

 39: Poison

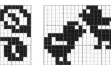 42: Cherries

 43: Chicks

 44: Lobster

 45: Bulb

 46: Cactus

 47: Kiwi

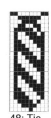 48: Tie

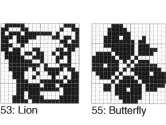 50: Falling Leaf

51: Boat

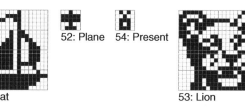 52: Plane

54: Present

 53: Lion

55: Butterfly

 49: Electric Guitar

56: Hand Grenade

58: Car

57: Dragon

59: Ship

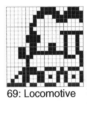
60: Tony

61: Wile-E Coyote

62: Spider

63: Cat

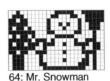
64: Mr. Snowman

65: Tractor

66: Tweet

67: Koala

68: Daffodil

69: Locomotive

70: Music Playing

71: Birds Gossip

72: Space Invaders

73: Look into My Eyes

74: Spider and Web

75: Eagle

77: I am Here!

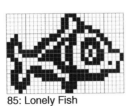
76: Pirate

78: Snake

79: Push the Trolley

81: Rabbit

80: Olive Branch

82: Turtle

83: Butterfly

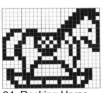
84: Rocking Horse

85: Lonely Fish

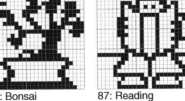
86: Bonsai

87: Reading

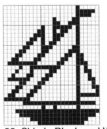
88: Ship in Black and White

89: Love

90: Water Mill

91: Flower Arrangement

92: Books D-G

93: Lighthouse

94: Flower

95: Octopus

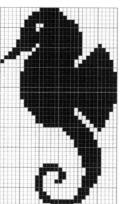

96: Do You Remember Furby? 97: Rose

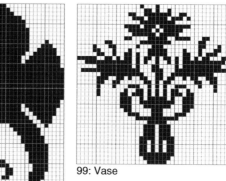

98: Seahorse

99: Vase

100: Fish

101: Scorpion

102: Cyclamens

103: Education

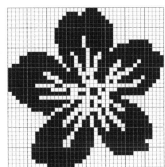
104: Flower Power

105: Heron

106: Serenade

107: Deer

108: Doll

109: Zebra

110: Mistletoe

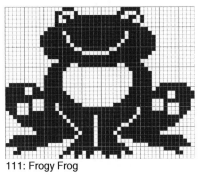
111: Frogy Frog

112: Reflection

113: Squirrel

114: Twins?

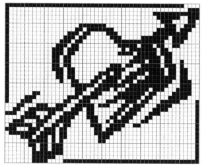

115: Love Struck Again

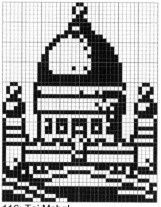

116: Taj Mahal

117: Lion

118: Pink Panther

119: Swan

120: Leave It to Him!

121: Playing the Lyre

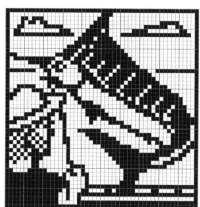

122: The Old Man and the Sea

123: In the Forest

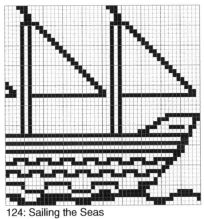

124: Sailing the Seas

125: "I Will Imprison You"

126: Cat

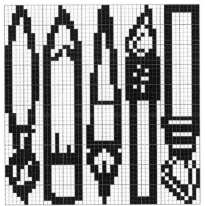

127: Art Supplies

128: Pagoda

129: Fish

130: Moby

131: Boy With Ball

132: Baby Bird

133: Lady

134: Leaf

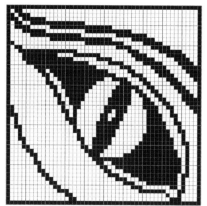
135: Dragon Eye

136: Guess Who?

137: Proposing

138: Creator

139: Giraffe

140: I Want to Dance

141: Cat

142: Bye Bye, Birdy

143: Aquarius

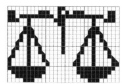
144: Libra

145: Virgo

146: Rhino

147: Slowly

148: Kayak Race

149: Time?

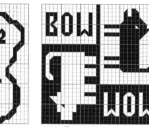
150: Bow-Wow

151: Fleeing, Clip Art

152: Mask

153: Paw Prints

154: Spider

156: Escher Triangle

155: Tinky Binky

157: Babyface

158: Iron

159: Video Camera

160: Mole

161: Man

162: Fishbowl

163: French Horn

griddlers
Logic Puzzles

Picture Logic Puzzles:

Griddlers

Griddlers are picture logic puzzles in which cells in a grid have to be colored or left blank according to numbers given at the side of the grid to reveal a hidden picture.

Triddlers

Triddlers are logic puzzles, similar to Griddlers, with the same basic rules of solving. In Triddlers the clues encircle the entire grid. The direction of the clues is horizontal, vertical, or diagonal.

MultiGriddlers

MultiGriddlers are large puzzles that consist of several parts of common griddlers. A Multi can have 2 to 100 parts. The parts are bundled and, once completed, create a bigger picture.

Word Search Puzzles:

Word Search

Word Search is a word game that is letters of a word in a grid. The goal of the game is to find and mark all the words hidden inside the grid. The words may appear horizontally, vertically or diagonally, from top to bottom or bottom to top, from left to right or right to left. A list of the hidden words is provided.

Each puzzle has some text and underscores (_ _ _) to indicate missing word(s). If the puzzle was solved successfully, the remaining letters pop up in the grid and the missing words appear in the text.

Smart Things Begin With Griddlers.net

griddlers
Logic Puzzles

Number Logic Puzzles:

8	1	2	5	7	9	4	3	6
4	6	9	1	3	2	5	8	7
5	3	7	6	8	4	9	1	2
6	7	4	8	5	1	2	9	3
1	8	3	2	9	7	6	4	5
2	9	5	4	6	3	1	7	8
9	5	8	3	1	6	7	2	4
7	2	6	9	4	8	3	5	1
3	4	1	7	2	5	8	6	9

Sudoku

Sudoku is a logic-based, number-placement puzzle. The goal is to fill a grid with digits so that each column and each row contain the digits only once.

4	2	3	6	5	1
1	5	4	3	6	2
6	1	2	5	3	4
2	3	6	4	1	5
3	4	5	1	2	6
5	6	1	2	4	3

Irregular Blocks (Jigsaws)

Jigsaw puzzle is played the same as Sudoku, except that the grid has Irregular Blocks, also known as cages.

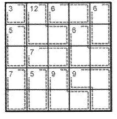

Killer Sudoku

The grid of the **Killer Sudoku** is covered by cages (groups of cells), marked with dotted outlines. Each cage encloses 2 or more cells. The top-left cell is labeled with a cage sum, which is the sum of all solution digits for the cells inside the cage.

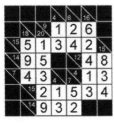

Kakuro

Kakuro is played on a grid of filled and barred cells, "black" and "white" respectively. The grid is divided into "entries" (lines of white cells) by the black cells. The black cells contain a slash from upper-left to lower-right and a number in one or both halves. These numbers are called "clues".

0	0	1	0	1	1	0	1
0	0	1	1	0	1	0	1
1	1	0	0	1	0	1	0
0	1	0	1	0	1	0	1
0	0	1	1	0	0	1	1
1	0	1	0	1	0	1	0
1	1	0	1	0	1	0	0
0	0	1	0	1	0	1	1

Binary

Complete the grid with zeros (0's) and ones (1's) until there are just as many zeros and ones in every row and every column.

Smart Things Begin With Griddlers.net

Number Logic Puzzles:

Greater Than / Less Than

Greater Than (or **Less Than**) Sudoku has no given clues (digits). Instead, there are "Greater Than" (>) or "Less Than" (<) signs between adjacent cells, which signify that the digit in one cell should be greater than or less than another.

Futoshiki

Futoshiki is played on a grid that may show some digits at the start. Additionally, there are "Greater Than" (>) or "Less Than" (<) signs between adjacent cells, which signify that the digit in one cell should be greater than or less than another.

Kalkudoku

The grid of the **Kalkudoku** is divided into heavily outlined cages (groups of cells). The numbers in the cells of each cage must produce a certain "target" number when combined using a specified mathematical operation (either addition, subtraction, multiplication or division).

Straights

Straights (Str8ts) is played on a grid that is partially divided by black cells into compartments. Compartments must contain a straight - a set of consecutive numbers - but in any order (for example: 2-1-3-4). There can also be white clues in black cells.

Skyscrapers

The **Skyscrapers** puzzle has numbers along the edge of the grid. Those numbers indicate the number of buildings which you would see from that direction if there was a series of skyscrapers with heights equal the entries in that row or column.

Smart Things Begin With Griddlers.net

Made in the USA
San Bernardino, CA
26 November 2016